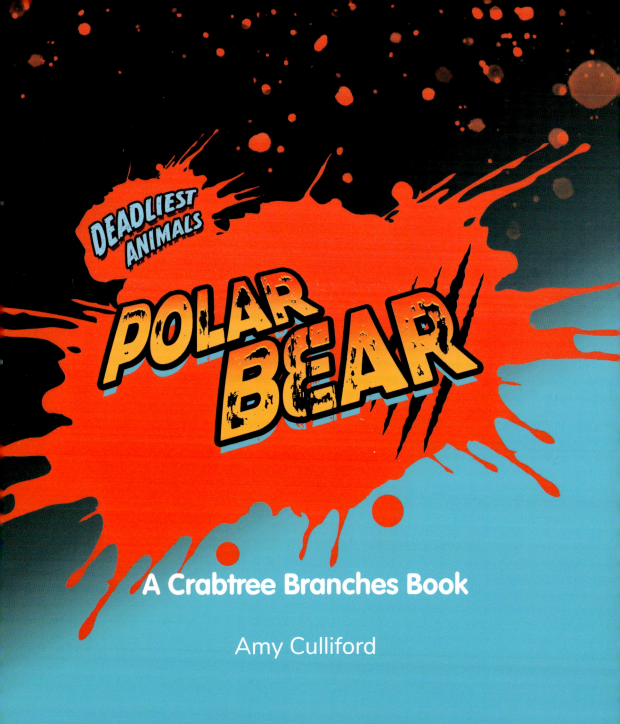

DEADLIEST ANIMALS
POLAR BEAR

A Crabtree Branches Book

Amy Culliford

CRABTREE Publishing Company
www.crabtreebooks.com

School-to-Home Support for Caregivers and Teachers

This high-interest book is designed to motivate striving students with engaging topics while building fluency, vocabulary, and an interest in reading. Here are a few questions and activities to help the reader build upon his or her comprehension skills.

Before Reading:
- *What do I think this book is about?*
- *What do I know about this topic?*
- *What do I want to learn about this topic?*
- *Why am I reading this book?*

During Reading:
- *I wonder why…*
- *I'm curious to know…*
- *How is this like something I already know?*
- *What have I learned so far?*

After Reading:
- *What was the author trying to teach me?*
- *What are some details?*
- *How did the photographs and captions help me understand more?*
- *Read the book again and look for the vocabulary words.*
- *What questions do I still have?*

Extension Activities:
- *What was your favorite part of the book? Write a paragraph on it.*
- *Draw a picture of your favorite thing you learned from the book.*

TABLE OF CONTENTS

Polar Bears 4
Arctic Home 6
Polar Bear Size 8
Polar Bear Fur 10
Family Life.. 12
Deadly Weapons.. 14
On the Hunt... 20
Shrinking Habitat... 24
Bear Attack!... 26
Glossary 30
Index 31
Websites to Visit.. 31
About the Author... 32

POLAR BEARS

Polar bears, sometimes called sea bears or white bears, are among the largest of all bears. They are **apex predators**, which means they have no true natural enemies.

Polar bears are the most **carnivorous** of all bears. They eat almost nothing but meat. Hungry polar bears are aggressive. This makes them very dangerous.

Polar bears are marine mammals. They spend much of their lives hunting seals on the sea ice.

ARCTIC HOME

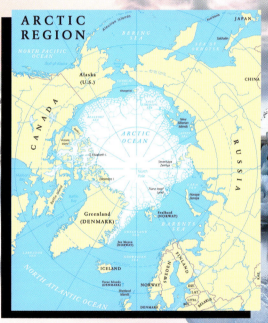

Polar bears are closely related to brown bears. But, unlike brown bears that roam woodlands, polar bears live on the sea ice. Their natural **habitat** is the Arctic— the region around the North Pole. They live in Greenland, Norway, Russia, Alaska in the US, and in Canada.

Polar bears evolved from brown bears hundreds of thousands of years ago. Over time, they adapted for life in the frozen Arctic.

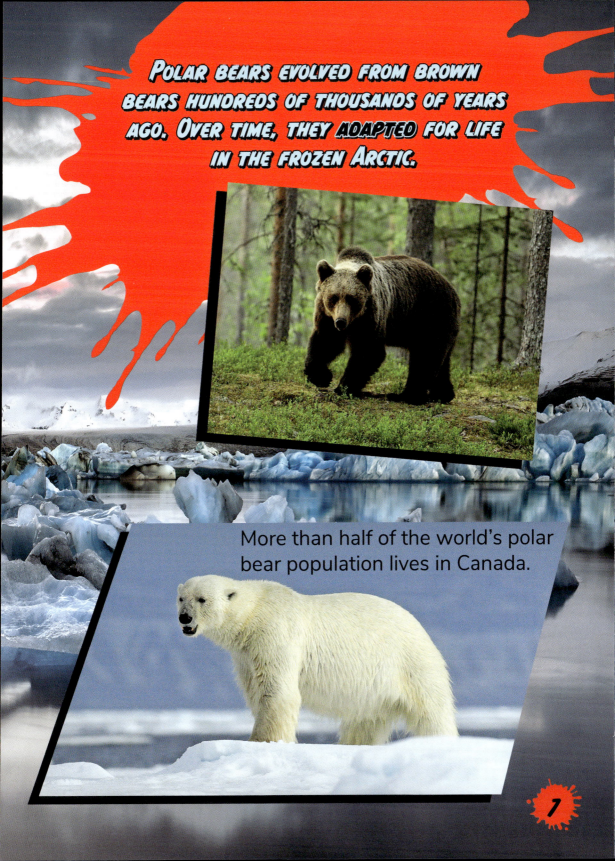

More than half of the world's polar bear population lives in Canada.

POLAR BEAR SIZE

Polar bears are massive. Adult females, called sows, can grow 8 feet (2.4 meters) in length and can weigh more than 600 pounds (272 kg).

Males, called boars, can weigh over 1,200 pounds (544 kg). They measure more than 10 feet (3 meters) in length. Imagine how a large male polar bear would tower over a human when it stands on its hind legs!

Polar bears live about 18 years in the wild, and about 30 years in captivity.

POLAR BEAR FUR

Polar bear fur appears white, off-white, and sometimes yellow. But, their fur is actually a mass of clear, hollow tubes. The white appearance comes from the way sunlight scatters in the tubes. Their fur is an **adaptation** that helps them thrive in their snowy home.

Two layers of fur keep polar bears warm in frigid air. Black skin underneath absorbs the sun's rays. A thick layer of fat beneath their skin keeps them warm in icy cold water.

FAMILY LIFE

Polar bears prefer **solitary** lives. However, males and females do come together to **mate**. After about eight months, pregnant females typically give birth to two cubs. They weigh about one pound at birth and are smaller than a loaf of bread.

Mother's milk helps cubs gain weight quickly. Cubs stay with their moms for about two years. Moms protect their cubs until they can fend for themselves.

Before giving birth, a female polar bear will dig a den in the snow. The den provides a safe place for a new mother and her cubs.

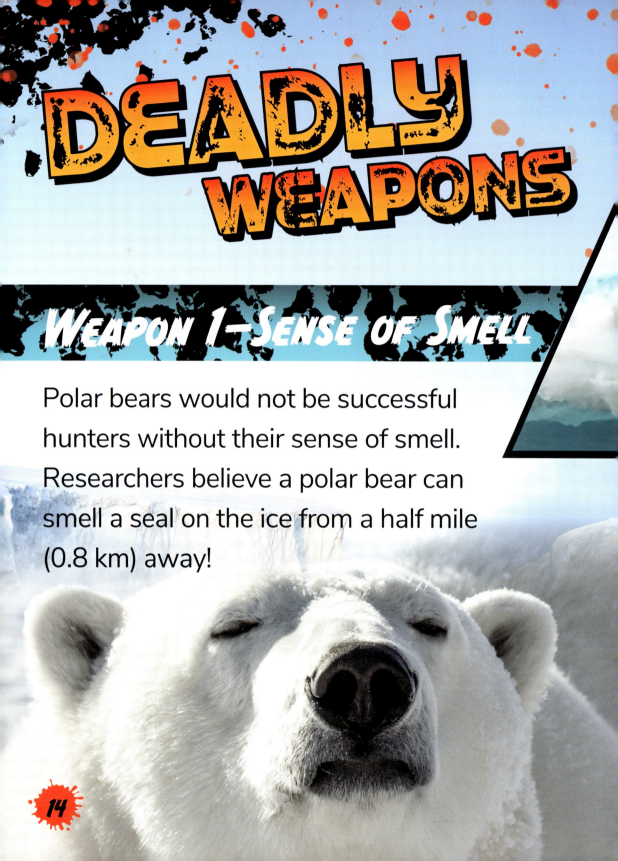

DEADLY WEAPONS

Weapon 1—Sense of Smell

Polar bears would not be successful hunters without their sense of smell. Researchers believe a polar bear can smell a seal on the ice from a half mile (0.8 km) away!

Polar bear vision is similar to ours, but their eyes have a membrane that protects them from harmful ultraviolet light.

Weapon 2—Paws, Claws, and Jaws

Polar bear paws are nearly 12 inches (31 cm) across! The huge paws help distribute weight evenly on thin ice. They are also ideal for propelling them through water.

Polar bears have razor-sharp claws that can rip through the skin of their **prey**.

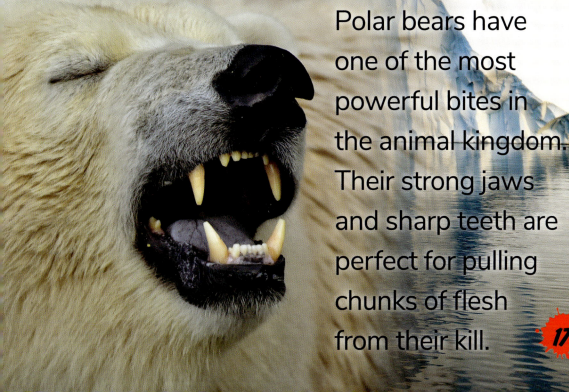

Polar bears have one of the most powerful bites in the animal kingdom. Their strong jaws and sharp teeth are perfect for pulling chunks of flesh from their kill.

Weapon 3 — Color

Polar bear fur is a type of **camouflage**. It blends in with snow. This makes it easier for them to sneak up on a resting seal.

Weapon 4 – Strength

Strong forearms and hind legs allow a hunting bear to grab a seal from an ice hole and drag it to a safe place to feed.

Polar bears are great swimmers. They can easily swim from one ice floe to another while traveling.

ON THE HUNT

Seals are a polar bear's favorite food. Polar bears use a strategy called still-hunting when searching for seals. A still-hunting bear uses its sense of smell to detect seal activity around openings in the ice.

A resting bearded seal must stay alert when polar bears are near.

The bear will wait close to the opening, sometimes for hours, for a seal to **breach**. When it does, the bear grabs the seal and drags it onto the ice.

Stalking is another polar bear hunting **strategy**. When a bear smells a seal resting on the ice, it begins to stalk. Its white camouflage helps the bear sneak up close. When the time is right, the bear charges and grabs the seal before it can slip back into the water.

Polar bears do not hibernate like brown bears. They must continue to hunt in the harsh winter months to sustain their energy and body weight.

SHRINKING HABITAT

The fatter a polar bear gets in winter, the better its chance of survival in the summer. That's because melting sea ice in the summer makes hunting more difficult. Unfortunately, climate change is causing sea ice to melt, or shrink, even in the winter months.

Polar bears depend on the sea ice for survival. As their natural habitat shrinks, their search for food takes them closer to human populations. This can be tragic for both bears and for people.

Polar bears that roam too close to human populations are darted and removed for the safety of the bear and people.

BEAR ATTACK!

In 2011, a group of students and guides were exploring in Norway. Being in polar bear country, they took precautions. The group had a gun, and for added protection, they ran a tripwire around their camp. If a bear tripped the wire, it would trigger a flare to warn the group. Unfortunately, the

MANY PEOPLE STILL TRAVEL, EXPLORE, AND CAMP ON THE SVALBARD ARCHIPELAGO WHERE THE BEAR ATTACK OCCURRED.

tripwire failed. During the night a hungry polar bear walked into camp, undetected. The bear ripped into one of the tents and attacked several men. A group leader managed to shoot and kill the bear. Sadly, it was too late. The bear killed one young man and injured several others.

Polar bears are the world's most dangerous bears. Fortunately, their interactions with people are not common. They live where human populations are few and far between. Polar bears are aggressive and have no fear of humans. Given the opportunity, hungry polar bears will hunt people. That's why polar bears are one of the deadliest animals on Earth.

Scientists can safely take measurements of a polar bear after it has been darted with a drug that makes it sleep.

Glossary

adaptation (ad-ap-TAY-shuhn): A change to fit in better

adapted (uh-DAPT-uhd): Changed to meet new conditions

apex predators (AY-peks PRED-uh-torz): Animals at the highest of the food chain

breach (BREECH): To come to the surface to breathe

camouflage (KAM-uh-flahzh): Coloring that makes animals look like their surroundings

carnivorous (kar-NIV-ur-uhss): Eating meat

habitat (HAB-i-tat): Natural place to live

ice floe (EYESS-FLO): Sheet of floating ice

mate (MAYT): To join for breeding

prey (PRAY): An animal that is hunted by another animal for food

solitary (SOL-uh-ter-ee): Being alone

strategy (STRAT-uh-jee): A plan or method created to achieve a goal

Index

Arctic 6, 7
cubs 12, 13
fur 10, 11, 18
habitat 6, 25
hunt(ing) 5, 14, 19, 20, 22, 24, 28
prey 16
seal 5, 14, 18, 19, 20, 21, 22
size 8-9

Websites to Visit

https://kids.nationalgeographic.com/animals/mammals/polar-bear/

https://kids.sandiegozoo.org/animals/polar-bear

https://polarbearsinternational.org/education-center

About the Author

Amy Culliford

Amy Culliford has a Bachelor of Fine Arts. She has worked as a classroom drama teacher and led after-school drama programs. She avoids bears of any kind.

The author would like to thank David and Patricia Armentrout for their research and help on this project.

Produced by: Blue Door Education for Crabtree Publishing
Written by: Amy Culliford
Designed by: Jennifer Dydyk
Edited by: Tracy Nelson Maurer
Proofreader: Crystal Sikkens

Photographs: Cover photo © Krasula/Shutterstock.com, graphic splat on cover and throughout © Andrii Symonenko /Shutterstock.com, page 4 © Ondrej Prosicky/Shutterstock.com, page 5 background photo © muratellioglu, bear © Alexey Seafarer/Shutterstock.com, pages 6-7 background photo © Jan Miko/Shutterstock.com, map © Peter Hermes Furian/Shutterstock.com, page 7 (top) © Erik Mandre/Shutterstock.com, (bottom) © ndrej Prosicky/Shutterstock.com, pages 8-9 background photo © ginger_polina_bublik/Shutterstock.com, page 8 © Olga_i/Shutterstock.com, page 9 (man) © Michal Sanca/Shutterstock.com, bear © evaurban/Shutterstock.com, page 10 © Ondrej Prosicky/Shutterstock.com, page 11 (top) © Mikhail Kolesnikov/Shutterstock.com, (bottom) © breakermaximus/Shutterstock.com, pages 12-19 background photo © muratellioglu/Shutterstock.com, page 12 © Alexey Seafarer/Shutterstock.com, page 13 (top) © jolly_photo/Shutterstock.com, (bottom) © Sergey Uryadnikov/Shutterstock.com, page 14 © Green Mountain Exposure, page 15 (top) © FloridaStock/Shutterstock.com, (bottom) © Mikhail Kolesnikov/Shutterstock.com, page 16 (top) © A. Laengauer/Shutterstock.com, bottom © KARI K/Shutterstock.com, page 17 (top) © Stefan Redel/Shutterstock.com, (bottom) © Unicorn555/Shutterstock.com, page 18 © jo Crebbin/Shutterstock.com, page 19 (top) © miroslav chytil/Shutterstock.com, (bottom) © Maximillian cabinet/Shutterstock.com, page 20 © Geoff Vago/Shutterstock.com, page 21 © Ondrej Prosicky/Shutterstock.com, pages 22-23 background photo © Jan Miko/Shutterstock.com, page 22 © GTW/Shutterstock.com, page 23 © La Nau de Fotografia/Shutterstock.com, pages 24-25 © FloridaStock/Shutterstock.com, inset photo © Andreas Weith https://creativecommons.org/licenses/by-sa/4.0/deed.en, pages 26-27 © Toranote/Shutterstock.com, pages 28-29 background photo © ginger_polina_bublik, page 28 © Thomas Barrat/Shutterstock.com, page 29 (top) © Igor Batenev/Shutterstock.com, (bottom) © Sergey Uryadnikov/Shutterstock.com

---

Library and Archives Canada Cataloguing in Publication

Title: Polar bear / Amy Culliford.
Names: Culliford, Amy, 1992- author.
Description: Series statement: Deadliest animals | "A Crabtree branches book". | Includes index.
Identifiers: Canadiana (print) 20210218940 |
Canadiana (ebook) 20210218967 |
ISBN 9781427154156 (hardcover) |
ISBN 9781427154217 (softcover) |
ISBN 9781427154279 (HTML) |
ISBN 9781427154330 (EPUB) |
ISBN 9781427154392 (read-along ebook)
Subjects: LCSH: Polar bear—Juvenile literature.
Classification: LCC QL737.C27 C85 2022 | DDC j599.786—dc23

Library of Congress Cataloging-in-Publication Data

Names: Culliford, Amy, 1992- author.
Title: Polar bear / Amy Culliford.
Description: New York : Crabtree Publishing, 2022. | Series: Deadliest animals - a Crabtree branches book | Includes index.
Identifiers: LCCN 2021022076 (print) | LCCN 2021022077 (ebook) |
ISBN 9781427154156 (hardcover) |
ISBN 9781427154217 (paperback) |
ISBN 9781427154279 (ebook) |
ISBN 9781427154330 (epub) |
ISBN 9781427154392
Subjects: LCSH: Polar bear--Juvenile literature. | Dangerous animals--Juvenile literature.
Classification: LCC QL737.C27 C749 2022 (print) | LCC QL737.C27 (ebook) | DDC 599.786--dc23
LC record available at https://lccn.loc.gov/2021022076
LC ebook record available at https://lccn.loc.gov/2021022077

Crabtree Publishing Company

www.crabtreebooks.com   1-800-387-7650

Copyright © 2022 **CRABTREE PUBLISHING COMPANY**

All rights reserved. No part of this publication may be reproduced, stored in a retrieval system or be transmitted in any form or by any means, electronic, mechanical, photocopying, recording, or otherwise, without the prior written permission of Crabtree Publishing Company. In Canada: We acknowledge the financial support of the Government of Canada through the Canada Book Fund for our publishing activities.

Published in the United States
Crabtree Publishing
347 Fifth Avenue, Suite 1402-145
New York, NY, 10016

Published in Canada
Crabtree Publishing
616 Welland Ave.
St. Catharines, ON, L2M 5V6

Printed in the U.S.A./062021/CG20210401